Changes in Matter

Copyright © by Harcourt, Inc.

All rights reserved. No part of this publication may be reproduced or transmitted in any form or by any means, electronic or mechanical, including photocopy, recording, or any information storage and retrieval system, without permission in writing from the publisher.

Requests for permission to make copies of any part of the work should be addressed to School Permissions and Copyrights, Harcourt, Inc., 6277 Sea Harbor Drive, Orlando, FL 32887-6777. Fax: 407-345-2418.

HARCOURT and the Harcourt Logo are registered trademarks of Harcourt, Inc., registered in the United States of America and/or other jurisdictions.

Printed in Mexico

ISBN 978-0-15-362047-8
ISBN 0-15-362047-1

2 3 4 5 6 7 8 9 10   805   16 15 14 13 12 11 10 09 08

Visit *The Learning Site!*
www.harcourtschool.com

Lesson 1

What Is Matter Made Of?

VOCABULARY
matter
atom
element

Matter is anything that has mass and takes up space. Matter makes up objects all around you.

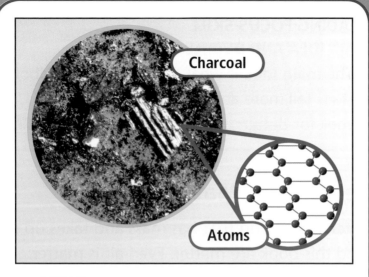

An **atom** is the smallest possible part of something that can exist. Atoms are so small that you need a special microscope to see them. These are charcoal atoms.

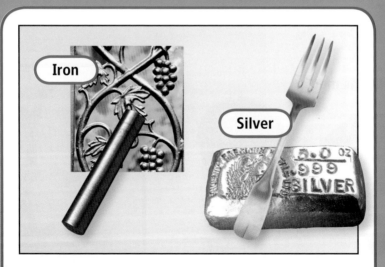

An **element** is a substance made up of only one kind of atom. Oxygen, carbon, mercury, iron, and silver are examples of elements.

READING FOCUS SKILL
MAIN IDEA AND DETAILS

The main idea is what the text is mostly about.
Details tell more about the main idea.
Look for details about matter and what it is made of.

Properties of Matter

Matter is anything that has mass and takes up space. You and this book are matter. Even air is matter.

Look at the soccer balls. One ball is filled with air. The other one is not. The balance shows that the ball with air has more mass. The extra mass comes from the air inside it.

 Give three examples of matter. How do you know each one is matter?

▼ Air is matter.

Toys are matter. ▶

Particles of Matter

All matter is made up of tiny particles, or bits. Each kind of matter is made up of certain kinds of particles.

Matter can be broken down, but only so far. An **atom** is the smallest part of an *element* that can exist. It is very small. You need a special microscope to see an atom.

 What is an atom?

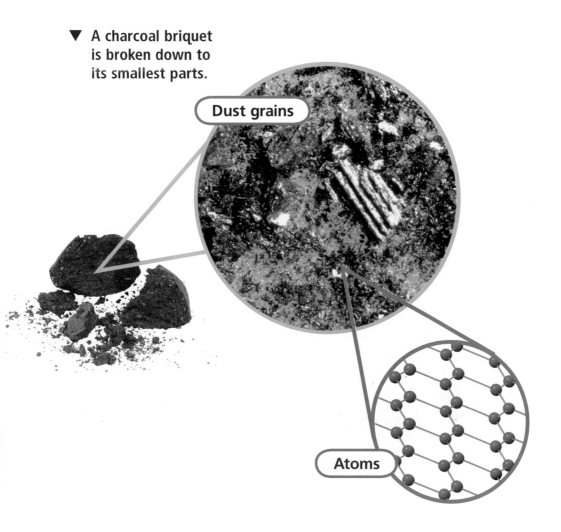

▼ A charcoal briquet is broken down to its smallest parts.

Dust grains

Atoms

Elements

An **element** is a substance made up of only one kind of atom. Oxygen is an element. It is made of atoms that are all oxygen. Carbon is another element. The atoms in charcoal and the tip of your pencil are carbon atoms. Silver, iron, mercury, and gold are elements, too. There are more than 100 elements.

 What is an element? Give two examples.

▼ Elements

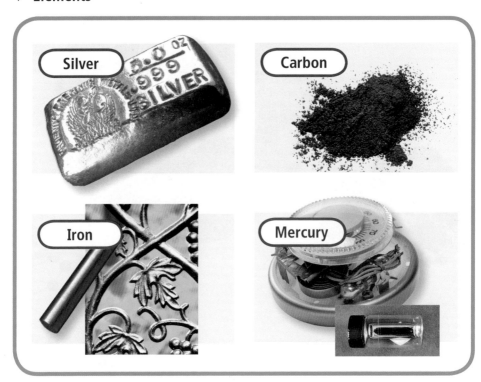

Groups of Elements

Elements are classified into groups. Two groups are metals and nonmetals. Gold, iron, and silver are metals.

Not all metals are elements. Steel is a metal but not an element. It is made of iron and carbon.

Most metals are shiny. They can be bent, rolled out, or made into thin wire. Non-metals, like sulfur, are dull and brittle. They break if you stretch them out.

 Name two groups of elements.

Gold ▶

▼ Sulfur

Review

Complete the main idea statement.

1. Matter is anything that has _____ and takes up _____.

Complete these detail sentences.

2. Tiny particles make up all _____.

3. An _____ is the smallest possible part of something that can exist.

4. Something that is made up of all the same kind of atoms is called an _____.

Lesson 2

VOCABULARY
change of state
physical change

What Are Physical Changes in Matter?

A **change of state** is when matter changes from one state to another. It can go from a solid to a liquid or from a liquid to a gas. These icicles are changing from a solid to a liquid.

A **physical change** is a change that does not make a new substance. Cutting paper is a physical change.

READING FOCUS SKILL
COMPARE AND CONTRAST

When you **compare and contrast** you tell how things are alike and different

Look for ways to **compare and contrast** states of matter.

States of Matter

Almost everything on Earth can be a solid, a liquid, or a gas. These are three states of matter. A **change of state** happens when matter changes from one state to another.

Adding or taking away heat changes the state of matter. For example, a solid melts into a liquid if it is heated enough. If heat is taken away from a liquid, it changes to a solid.

Liquid
Particles in liquid water are close together and move quickly.

The particles in matter are always moving. The difference between states of matter is how the particles move. For example, when water is ice, the particles are packed. They vibrate in place. As liquid water, the particles move easily and slide around. As water vapor, the particles move quickly. They are far apart.

 Compare the particles in ice, water, and water vapor.

Solid
Particles in ice are locked in place, but still moving.

Gas
Particles in gases, such as air, are far apart and move very quickly.

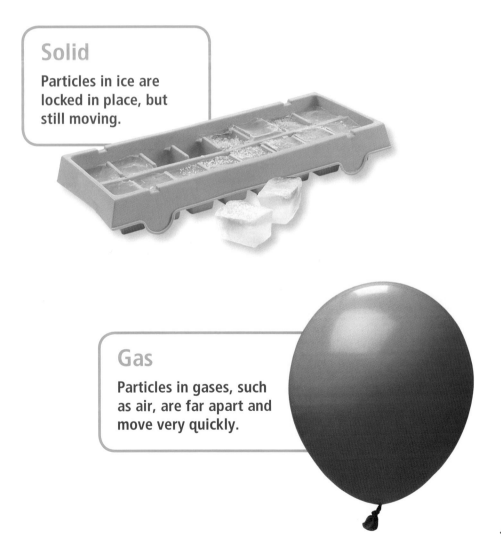

Physical Changes

You know that water can change states. When ice melts into water or water changes to water vapor, it is still water. It is just in a different form. A change of state is a physical change. A **physical change** is a change that does not make a new substance.

A change of state is a physical change. ▼

Icicles melting

Water boiling

Shredding

Cutting

Carving

What happens to paper if you shred it or cut it? The size or shape of the paper will change. But the pieces are still the same paper. What happens to a log if you carve it? The saw will make lots of chips. But the chips are still wood. Shredding, cutting, and carving are physical changes, too.

 How are all physical changes the same?

Dissolving

You know that a change of state is a physical change. *Dissolving* is another kind of physical change. If you mix sugar and hot water, the sugar dissolves. That means it mixes evenly into the water. Once the sugar dissolves, you can no longer see it.

▼ **Sugar dissolving in hot water**

You can show that dissolving is a physical change. When you let the water evaporate, sugar is left behind. It has not changed into new matter.

 Tell how dissolving is like cutting.

Water has evaporated. Sugar is left. ▶

Review

Complete the compare and contrast statements.

1. A solid changing to liquid and a liquid changing to gas are both changes of _____.

2. A liquid changes to a solid when heat is _____, and a solid changes to a liquid when heat is _____ to it.

3. Shredding, cutting, and dissolving are all kinds of _____ changes.

4. Liquids, solids, and gases are all made up of moving _____.

Lesson 3

How Does Matter React Chemically?

VOCABULARY

physical property
chemical property
chemical change
chemical reaction
compound

A **physical property** is something you can observe about matter. Color, texture, and size are physical properties.

A **chemical property** is information about how matter interacts with other matter. For example, a chemical property of iron makes it form rust when water is present.

In a **chemical change**, matter reacts with other matter to form new matter. **Chemical reaction** is another name for chemical change. Rusting is a chemical change.

A **compound** is a substance made up of two or more elements. It is the result of a chemical reaction. For example, water is a compound. It is made of hydrogen and oxygen atoms.

READING FOCUS SKILL
COMPARE AND CONTRAST

When you compare and contrast you tell how things are alike and different.

Compare chemical changes to physical changes.

Chemical and Physical Properties

How do you describe a pencil? You may say that it is long, thin, and yellow. These are physical properties. A **physical property** is information about matter that you can observe. You can see and measure physical properties.

		Physical Properties	
Water		• colorless • odorless • liquid at room temperature	• boils at 100°C • melts at 0°C
Silver		• shiny • soft • silver in color	• boils at 2163°C • melts at 962°C
Iron		• shiny • hard • grayish silver in color	• boils at 2861°C • melts at 1538°
Sulfur		• dull • brittle • yellow	• boils at 445°C • melts at 115°C

Look at the table. It lists physical properties of water, silver, iron, and sulfur.

You can describe chemical properties of matter, too. A **chemical property** is information about how matter interacts with other matter. Think about the pencil. It will burn if there is oxygen and the temperature is high enough. This tells about a chemical property of a pencil.

Look at the table. It lists chemical properties of water, silver, iron, and sulfur.

 Tell how a physical property and a chemical property are the same and different.

Compare the physical and chemical properties of water, silver, iron, and sulfur. ▼

Chemical Properties
• made up of hydrogen and oxygen • many substances dissolve easily in it
• does not react with many other substances • does not react with air • reacts with ozone or sulfur to form tarnish
• reacts easily with many other substances • reacts with oxygen to form the minerals hematite and magnetite • reacts with oxygen in presence of water to form rust
• reacts with any liquid element • reacts with any solid element except gold and platinum • reacts with oxygen to form sulfur dioxide, a form of air pollution

Chemical Changes

Hydrogen and oxygen are elements and gases, too. When hydrogen burns, it combines with oxygen to form water. This is a chemical change. In a **chemical change**, matter reacts with other matter to form new substances. **Chemical reaction** is another name for chemical change.

▼ Sulfur helps a match to light quickly.

▼ Iron reacts with oxygen and water to form rust.

Water is a compound. A **compound** is a substance made up of two or more elements. It is the result of a chemical change.

The pictures show other chemical changes.

 How is a chemical change different from a physical change?

▲ Silver reacts with sulfur in the air to form tarnish.

Recognizing Chemical Changes

How do you know when a chemical change is taking place? Your senses can help you. For example, before you bake a loaf of bread, you see that the dough is white. After it is in the oven a while, you begin to smell it baking. Once it is baked, you see that the crust is brown. The bread can never go back to the way it was before it was baked. These clues tell you that a chemical reaction has taken place.

Before ▶

After ▶

Look at the table. It can help you decide when a chemical change is taking place.

Try not to confuse a chemical change with a physical change. For example, when water freezes, it becomes a solid. It changes to a new state. A change of state is a physical change, not a chemical change.

 Tell how a baking loaf of bread and a rusting car are similar.

Chemical Changes ▼

Clues to Chemical Changes

Clue	Example	Description
Color Change	Bread dough baking	Changes from white to brown
Smell	Eggs rotting	Gives off a terrible smell
New Physical Property	Iron rusting	Changes from hard and silvery to brittle and reddish brown
Substance Given Off	Wood burning	Smoke is released into the air
Heat Given Off	Sulfur burning	Fire is hot

Review

Complete the compare and contrast statements.

1. Some properties are physical and some properties are _____.

2. Oxygen is an example of an _____ and water is an example of a _____.

3. Rotting eggs and baking bread are both _____ changes.

GLOSSARY

atom (AT•uhm) the smallest unit of an element that can exist

change of state (CHAYNJ uhv STAYT) a physical change that occurs when matter changes state such as from a liquid to a gas

chemical change (KEM•ih•kuhl CHAYNJ) a change in which matter reacts with other matter to form new substances

chemical property (KEM•ih•kuhl PRAHP•er•tee) information about how matter interacts with other matter

chemical reaction (KEM•ih•kuhl ree•AK•shuhn) a chemical change

compound (KAHM•pownd) a substance made of two or more elements

element (EL•uh•muhnt) a substance made up of only one kind of atom

matter (MAT•er) anything that has mass and takes up space

physical change (FIZ•ih•kuhl CHAYNJ) a change in matter that doesn't result in a different substance

physical property (FIZ•ih•kuhl PRAHP•er•tee) information about matter that you can observe